# WHAT IS MELITTOLOGY?

*Melittology* is the scientific study of bees.

The scientists who study bees are called **MELITTOLOGISTS**.

# ARE BEES NATURE'S TINY HEROES?

DISCOVER THE SCIENCE BEHIND **MELITTOLOGY**
(*meh-LIT-oh-luh-jee*)

Written by Eliza Jeffery
Illustrated by Lays Bittencourt

Words that are tricky to understand are in **bold**. Find out what they mean in the glossary.

Words that are difficult to say are in *italics*. Find out how to say them at the back of the book.

We simply wouldn't be able to survive without bees. Animals, insects, plants, and people all rely on them. Bees are the most important **pollinators** on planet Earth! Without them, it would be a very different world.

People who study bees are called *melittologists*, and understanding **pollination** is a vital part of their work. Bees may not know it, but by pollinating, they are keeping our natural world healthy.

**Pollen** sticks to them as they search for **nectar** and falls off when they visit the next flower. Wherever pollen lands, new plants, such as trees and flowers, grow!

Our **ecosystems** need bees! They support the whole **food chain.** By pollinating, bees support other insects, birds, and **mammals...**

including us! Bees help crops grow, which then feeds **livestock** and

**KEEPS TASTY FOOD ON OUR PLATES!**

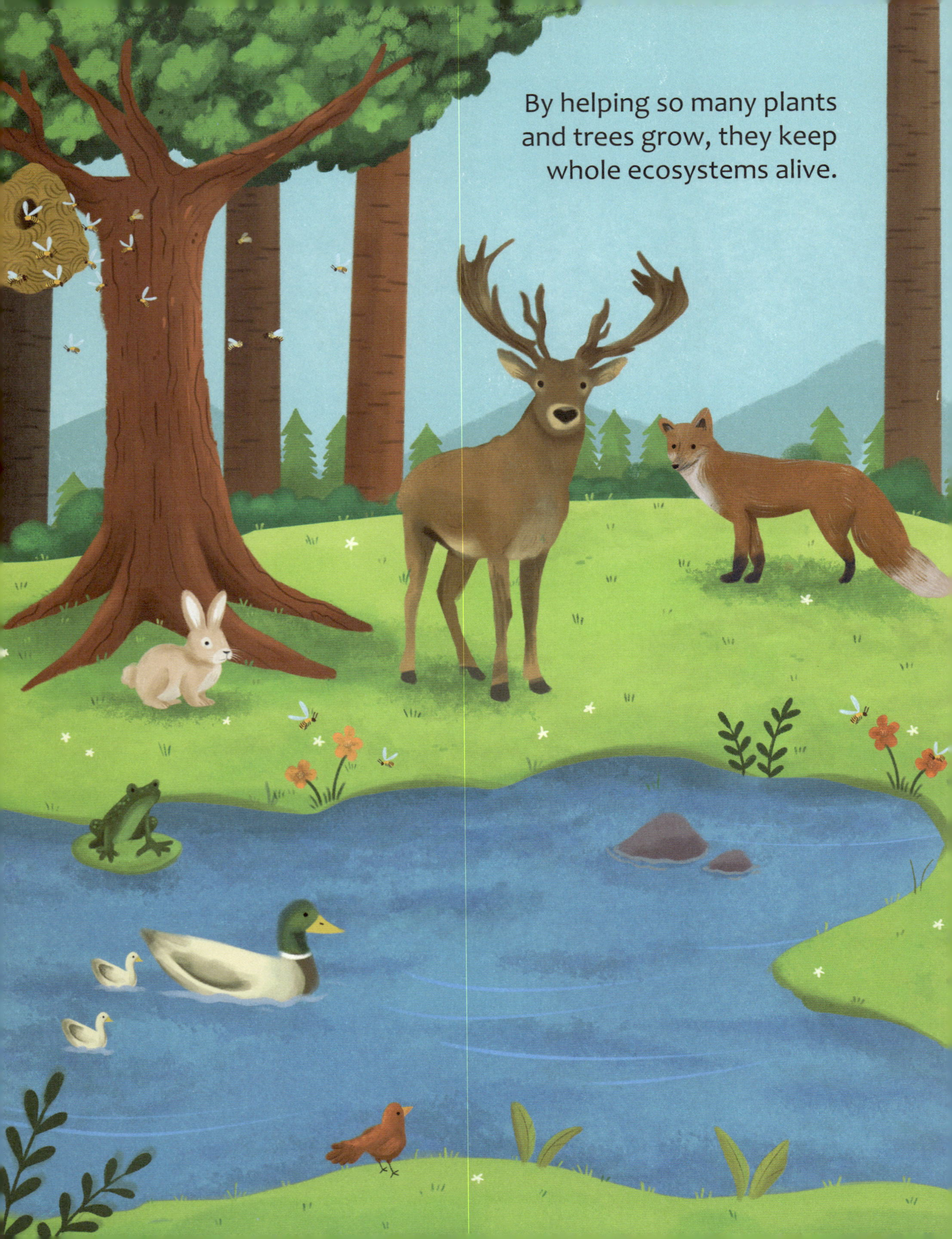

By helping so many plants and trees grow, they keep whole ecosystems alive.

A world without bees would look very different to our world now. There would be less food growing, **habitats** would disappear, and animals would have nowhere to live!

Honey is great to eat, but it is useful in other ways too. Ancient Egyptians used it to heal **wounds** and **embalm** the dead! Today, it's used for soothing sore throats, treating burns, and much more. Bees have been

**HELPING US FOR GENERATIONS!**

Bees make honey by drinking nectar from flowers! It is stored in their tummies and passed from bee to bee, where it turns into honey.

Bees then use the sugar in the honey to make **beeswax** – this is used to make the honeycomb! Bees store honey here for winter.

They use the honeycomb as a place for **worker bees** to live and grow, too. Melittologists are amazed by how these tiny creatures can make such perfect hexagonal shapes!

Some bees can sniff out the scent of other bees

**WITH THEIR FEET!**

Their "smelly" feet can tell that a flower has already been visited, so they waste no time and move on.

Their wings are awesome too. They can fly between 24 and 32 km per hour (15-20 mph). The speedy flapping of their wings makes the buzzing sound you can hear!

Bees live in large groups called **colonies** where each bee has an important role. The colony is ruled by a **queen bee.**

Some pollinate, some make honey, and others look after baby bees.

**IT TAKES TEAMWORK TO RUN A COLONY!**

If the colony is overcrowded, nectar is running low, or if they need to reproduce,

**BEES MAY SWARM!**

When a new queen bee emerges, **scout bees** find a place to build another home. The cloud of bees may look scary, but it's important for their survival.

Bees are facing big problems that are also hurting the rest of nature. These include **pollution** and **disease**...

high temperatures and **drought** caused by climate change...

and habitat loss. Melittologists work hard to protect bees all around the world, but they need all the help they can get.

BUG HOTEL
SEEDS

These problems may seem big, but we only need to make small changes to create a better world for bees! We could stop the use of **pesticides**, grow bee-friendly flowers, and provide shelter and water for them too.

Beekeepers also help to protect bees. They give them safe homes and keep them free from disease. Without beekeepers, there might not be enough bees to do all their clever jobs.

Bees may be tiny, but what they do for the world is huge! It is so important that we all work together to protect them.

Now we know bees are nature's tiny heroes. Melittologists have found so many different types of bees, and they're all saving our natural world, one little bee at a time!

Bumble bee

Honey bee

Carpenter bee

Common carder bee

Leaf-cutter bee

Mason bee

American bumble bee

Ivy bee

Blue orchard mason bee

Red-tailed bumble bee

Tawny mining bee

# The POLLINATION JOURNEY

**Bees are one of the most important pollinators in the world. Let's dive into how pollination works and why it matters for our natural world!**

### 1. *Collecting food*

Bees fly from their hive to find food like nectar and pollen in flowers.

### 2. *Gathering pollen*

When bees land on flowers, they pick up tiny grains called pollen on their bodies.

### 3. Moving pollen

As bees visit more flowers, some of the pollen they pick up rubs off onto other flowers.

### 4. Making seeds

Pollen that lands on a flower helps to form seeds that start to grow inside the flower.

### 5. Growing fruits

After this, the flower turns into a fruit, which protects the seeds as they grow.

### 6. Spreading seeds

When the fruits are ready, they release the seeds, which can grow into new plants elsewhere.

*Brilliant*

# BEE FACTS

**So we know how important bees are to our planet, but what else is there to know about these tiny creatures? What other facts have melittologists discovered about bees?**

## WAGGLE DANCE

Honey bees communicate with each other through a **unique** dance – by waggling their bodies! They do this to let other bees know where pollen can be found.

## A UNIQUE BUZZ

Different **species** of bees have unique buzzing sounds, and scientists can tell them apart just by listening!

## HOW MUCH HONEY?

Worker honey bees only live to be 7 weeks old! Despite all their hard work, these bees will only make enough honey to fill a tiny part of a teaspoon in their lifetime!

## HONEYCOMB HOME

Bees use honeycomb as a nursery as well as a honey-making factory! The queen lays eggs inside and they eventually turn into baby bees.

## TINY AND TIDY

Bees are very tidy creatures, often grooming each other and keeping their hives clean.

# GLOSSARY

**Beeswax** – a substance used by honey bees for building their hive.

**Colonies** – a group of the same type of creatures that live or grow together.

**Disease** – a condition that causes part of a living thing to no longer work properly.

**Drought** – a long period of dryness, usually caused by lack of rainfall.

**Ecosystems** – all the living and non-living things that exist together within an area.

**Embalm** – to preserve a dead body.

**Food chain** – the order in which different species eat each other to survive.

**Habitats** – the places where animals or plants live.

**Livestock** – domestic animals kept for food, milk, leather, or wool.

**Mammals** – warm-blooded animals (including humans) that produce milk to feed their young.

**Nectar** – a sugary liquid produced by flowers that attracts pollinators (see below).

**Pesticides** – substances made to kill weeds, insects, rodents, or bacteria.

**Pollen** – a fine powder produced by some plants when they reproduce.

**Pollination** – the process of animals helping plants to make seeds by taking pollen from one plant to another plant of the same type. *Need help saying this? Look below!*

**Pollinators** – animals and insects that help transfer the small pollen (see above) grains needed by plants to make seeds and grow into new plants.

**Pollution** – adding harmful materials into the environment.

**Queen bee** – the only female in the colony who lays eggs.

**Scout bees** – worker bees whose role is to find a new nest.

**Species** – a group of living things that share characteristics and features. For example, honey bees and bumble bees are different species.

**Unique** – something that is one of its kind; unlike all others.

**Worker bees** – bees that build nests, guard the hive, look for food, or tend to young bees.

**Wounds** – injuries to the skin.

# HOW DO I SAY?

**Melittologists**
meh-LIT-oh-luh-jists

**Melittology**
meh-LIT-oh-luh-jee

**Pollination**
pol-uh-NAY-shun

# THE BIG QUESTIONS ANSWERED

**This is more than just a series of books; it is a complete resource. Accompanying each book is a variety of FREE material to engage curious kids with science.**

**www.thebigquestionsanswered.com**

Use the QR code to visit the website, download free resources, and discover other books in the series.

On the website, find out incredible things about melittologists, including what they do, some of their greatest discoveries, and what it takes to become an expert in this field of science.

**The material is also available for home or classroom use, supporting all the information in this book.**

**Teachers' & Parents' Resources**
With discussion prompts and questions, extra information, and facts around key topics.

**Young Melittologists' Activity Pack**
Fun activities for wannabe bee experts, including creative writing, drawing, word searches, and much, much more.

BEETLE BOOKS

**Beetle Books is an imprint of Hungry Tomato Ltd.**

First published in 2024 by Hungry Tomato Ltd
F15, Old Bakery Studios, Blewetts Wharf, Malpas Road,
Truro, Cornwall, TR1 1QH, UK.

ISBN 9781835691311



A CIP catalog record for this book is available from the British Library.

With thanks to:
Editor: Millie Burdett
Editor: Holly Thornton
Senior Designer: Amy Harvey
Tim Cook for his valued contribution
The team at Beehive Illustration

Printed and bound in China.

Picture Credits:
(t = top, b = bottom, m = middle, l = left, r = right)
Shutterstock: BikG 35tl; Kuttelvaserova Stuchelova 34bl; MeggedyannePhotography 34mr; Nicemyphoto 35bl; Tools Konten 35mr.